Next Steps for TIMSS

Directions for Secondary Analysis

Alexandra Beatty, Lynn W. Paine, and Francisco O. Ramirez, *Editors*

Board on International Comparative Studies in Education
Board on Testing and Assessment

Commission on Behavioral and Social Sciences and Education

National Research Council

NATIONAL ACADEMY PRESS
Washington, DC 1999

NATIONAL ACADEMY PRESS • 2101 Constitution Avenue, N.W. • Washington, DC 20418

The study was supported by Grant No. ESI-9355774 between the National Academy of Sciences and the National Science Foundation. Any opinions, findings, conclusions, or recommendations expressed in this publication are those of the author(s) and do not necessarily reflect the view of the organizations or agencies that provided support for this project.

International Standard Book Number 0-309-06428-7

Additional copies of this report are available from National Academy Press, 2101 Constitution Avenue, N.W., Washington, DC 20418

Call (800) 624-6242 or (202) 334-3313 (in the Washington metropolitan area)

This report is also available on line at **http://www.nap.edu**

Printed in the United States of America

BOARD ON INTERNATIONAL COMPARATIVE STUDIES IN EDUCATION

Contents

Preface

The Board on International Comparative Studies in Education (BICSE) was established by the National Research Council (NRC) in 1988 at the request of the U.S. Department of Education's National Center for Education Statistics and the National Science Foundation. The board's mandate over the past nine years has been to monitor the overall quality of international comparative studies, and to foster understanding about how such comparisons can be developed and used. Not surprisingly, the board has devoted considerable energy to monitoring each phase of The Third International Mathematics and Science Study (TIMSS), a study conducted by the International Association for the Evaluation of Educational Achievement. BICSE has followed debates about conceptual, technical, and management issues and has offered both counsel and a forum for discussion as the study has proceeded. In early 1997, as the first study results were released, the board cosponsored (with several other NRC bodies) a symposium at which leading scholars, policy makers, and TIMSS researchers reflected on the first set of results, methodological issues raised by the study, and the study's implications for education reform (see *Learning from TIMSS: Results of the Third International Mathematics and Science Study*, National Research Council, 1997).

Now that the initial results have been released, the board has turned its attention to what happens next. The TIMSS data are potentially useful to researchers, policy makers, practitioners, and others interested in evidence regarding factors that influence student learning. But although the study has produced a remarkable volume of intriguing data, it is by no means complete. Scholarly review of the initial data, evaluations of claims based on the data, and follow-up secondary analysis based on the primary findings are all integral parts of a study of this magnitude, but the bulk of this very important work has not yet begun. Because of the board's serious concern that this necessary work has not been undertaken, or funded, it held a workshop on June 17 and 18, 1998, to explore different perspectives on possible next steps.

The workshop was an invaluable opportunity for the board to explore issues and questions it has addressed over the years and to solidify its thinking about many of them. Because the board is convinced of the importance of moving forward with the TIMSS data, it presents in this report both recommendations as to what ought to be done and many of the innovative specific ideas that emerged from the workshop. These recommendations reflect the board's conviction, based on its many years of involvement with and deliberations about TIMSS, that this study is an extremely rich resource for the policy, scholarly, and practice communities, and that all of these groups have a responsibility to take full advantage of it. The recommendations and discussion in this report are intended to assist both researchers and funders who are considering further work with TIMSS, and a broader audience of researchers, policy makers, practitioners, and others who have followed the TIMSS results and are eager to use them. This report is, in a sense, the culmination of many years of effort for the board.

On behalf of the entire board, I extend particular thanks to a number of people whose help in this undertaking has been invaluable. Two board members, Lynn Paine and Francisco Ramirez, took the lead in framing the issues the board needed to address and in planning the workshop. They devised a workshop format that would make stimulating interactions among a diverse group, and on a complex set of issues, possible. It was challenging for the planners and participants alike, but the results were well worth it. Two other board members, Andrew Porter and John Dossey, also played key roles in the planning and in the workshop itself, leading discussions and providing support throughout the process. The board also owes a debt of gratitude to Harry Judge and Stephen Heynemann, who, by serving as rapporteurs at the workshop, took on the challenge of helping the board to synthesize the discussion. Support from staff members Patricia Morison, who has guided the board throughout the process, and Alix Beatty, who assisted with the workshop planning and drafted this document, is also gratefully acknowledged. Jane Phillips provided very able administrative support throughout the workshop, as she does for all BICSE activities.

BICSE carries out its work under the auspices of the NRC's Board on Testing and Assessment and several of its members have assisted in shaping the workshop and report. BICSE is particularly grateful to Richard Shavelson, former chair of BOTA, for his help at several stages of the process. Thanks are also due to our sponsors, NCES and NSF, for their support during the planning of the workshop. In particular, Eugene Owen at NCES and Larry Suter at NSF have been instrumental in facilitating BICSE's involvement in important issues throughout the design and execution of TIMSS.

The key to the success of this board project, however, lay in the contributions of 13 scholars who grounded the discussion with thoughtfully written responses to questions the board had framed. The board extends its deep appreciation to David Baker, Douglas Grouws, Jeremy Kilpatrick, Elizabeth King, Gerald LeTendre, Mary Metz, Heinrich Mintrop, Richard Murnane, Gary Natriello, Aaron Pallas, Senta Raizen, Mavis Sanders, and James Shymansky. By supplying their responses in advance of the workshop so that all participants could read them before convening, these scholars made it possible for the group to plunge quickly into substantive discussions of the issues about which the board was concerned. This diverse group also broadened the discussion and provided countless reminders of the many perspectives that need to be considered. The board asked a number of individuals who had not previously studied the TIMSS results to take on this task, and particularly appreciates their willingness to review a large volume of material in preparation for the workshop.

I also wish to thank all of the board members for the thoughtful ideas and stimulating discussions that led to the workshop and the production of this document. Their persistence in both tracking the complexities of TIMSS and thinking carefully about its implications over the years has yielded many valuable contributions to the process.

This report has been reviewed in draft form by individuals chosen for their diverse perspectives and technical expertise, in accordance with procedures approved by the NRC's Report Review Committee. The purpose of this independent review is to provide candid and critical comments that will assist the institution in making the published report as sound as possible and to ensure that the report meets institutional standards for objectivity, evidence, and responsiveness to the study charge. The review comments and draft manuscript remain confidential to protect the integrity of the deliberative process.

We wish to thank the following individuals for their participation in the review of this report: Paul J. Black, School of Education, King's College, London; Andrea A. diSessa, Graduate School of Education, University of California, Berkeley; Edward Hiler, vice chancellor for agriculture and life sciences, Texas A&M University; Michael Martinez, Education Department, University of Calfornia, Irvine; Stephen W. Raudenbush, Department of Counseling, Educational Psychology, and Special Education, Michigan State University; Catherine Snow, Graduate School of Education, Harvard University; and P. Michael Timpane, RAND, Washington, D.C. While the individuals listed above have provided constructive comments and suggestions, it must be emphasized that responsibility for the final content of this report rests entirely with the authoring board and the institution.

Michael W. Kirst, *Chair*
Board on International Comparative
Studies in Education

Next Steps for TIMSS

Directions for Secondary Analysis

Introduction

OVERVIEW

International comparisons have become a key element in the public discussion of how the U.S. educational system is functioning and what needs to be done to reform it. Curiosity about such comparisons may have begun with the urge to match or exceed the performance of our economic rivals, but it has extended to a much broader range of issues. The importance of cross-national comparisons of virtually all aspects of the delivery of education—curriculum, resources, governance and structure, and teacher development, to name a few—is now widely recognized. The Third International Mathematics and Science Study (TIMSS), one in a series of large-scale international assessments that have been conducted by the International Association for the Evaluation of Educational Achievement (IEA), has played an important role in illustrating the potential value of such comparisons.

TIMSS was a bold step forward not only for the relatively small community of scholars who have been particularly interested in international comparisons of education, but also for the much broader community of scholars, policy makers, teachers, and others concerned with education generally. The study was large—nearly 50 countries and more than half a million students participated—and complex. The list of U.S. publications devoted just to reporting the initial results stands at fourteen and counting.[1] But the study's boldness lay not just in its scale—its mission was extremely ambitious as well. Those who planned and executed the study intended not only to overcome the problems with sampling and other technical issues that plagued previous international studies, but also to collect a range of data sufficient to yield powerful insights into the contexts in which learning takes place around the world, as well as to identify compelling links between specific factors and high achievement (see Bracey, 1996; Rotberg, 1990).

The combination of generally recognized high technical quality and ambitious goals has brought the TIMSS results considerable attention in the United States at a time when education issues have been a major political focal point.[2] TIMSS findings have been made widely available through the production of a "Resource Kit" for districts and administrators, Internet discussion links and websites, and

[1]Appendix A is a "road map" to TIMSS that provides a brief description of each of its components, the data available, and sources for further information, as well as titles of the TIMSS reports published to date.

[2]Questions that have been raised about the technical quality of TIMSS are addressed later in this document.

a variety of other means.[3] TIMSS findings have been cited in political discourse from the presidential level to the school-board level. The desire for data that could be used to influence policy decisions was clearly a prime reason the United States chose to participate in the study and to commit funds to cover not only the costs of collecting U.S. data, but also a significant portion of the costs of collecting much of the international data, and the costs of the three-country (United States, Japan, and Germany) qualitative studies that helped put the achievement findings in context. However, the very richness and complexity of the study has been a source of dissonance between the research and policy communities. Although policy makers are eager for information they can readily use and apply in planning and decision making, many of those who know TIMSS best are leery of the oversimplification and misinterpretation that can creep in when quick answers are sought.

As with any study of its kind, the value of TIMSS will lie not just in the official reports of the data collected, but also in the efforts of the many researchers—not part of the original TIMSS team—who are expected to conduct secondary analyses, using the initial data as the basis for further explorations of specific questions. TIMSS was deliberately designed to make use of a variety of methodologies so that questions about the factors that influence young people's learning could be examined using a variety of types of data. Figure 1 illustrates the basic structure of TIMSS. Although the conceptual framework on which the study was based called for the various components of the study—the studies of achievement and curriculum and the video and case studies—to be linked together, there is little direct precedent for doing so on the scale of TIMSS. Moreover, exploring such links was not the principal focus of the researchers who produced the first wave of reports from TIMSS, since it was necessary to complete each of the parts before thoughtful links among them could be attempted. However, the members of the Board on International Comparative Studies in Education (BICSE) have perceived, in public discussion of the published reports from TIMSS, a growing sense that the study is complete and that the only work left to be done is translation of the results already released into guidelines for practice and policy. What has been largely ignored in this discussion is the important challenge of establishing links among the study's components. The board believes strongly that a great deal more work needs to be done with TIMSS, and that without further effort, extremely valuable opportunities to learn from an intensive, and expensive, data collection effort are in danger of being lost. The board is not concerned only about the scholarly community's pursuit of intellectual issues arising from TIMSS, though those are important. Of more immediate concern for the board are two issues: first, that much-needed analyses have not

[3]See Appendix A for information about the Resource Kit, websites, and other TIMSS resources.

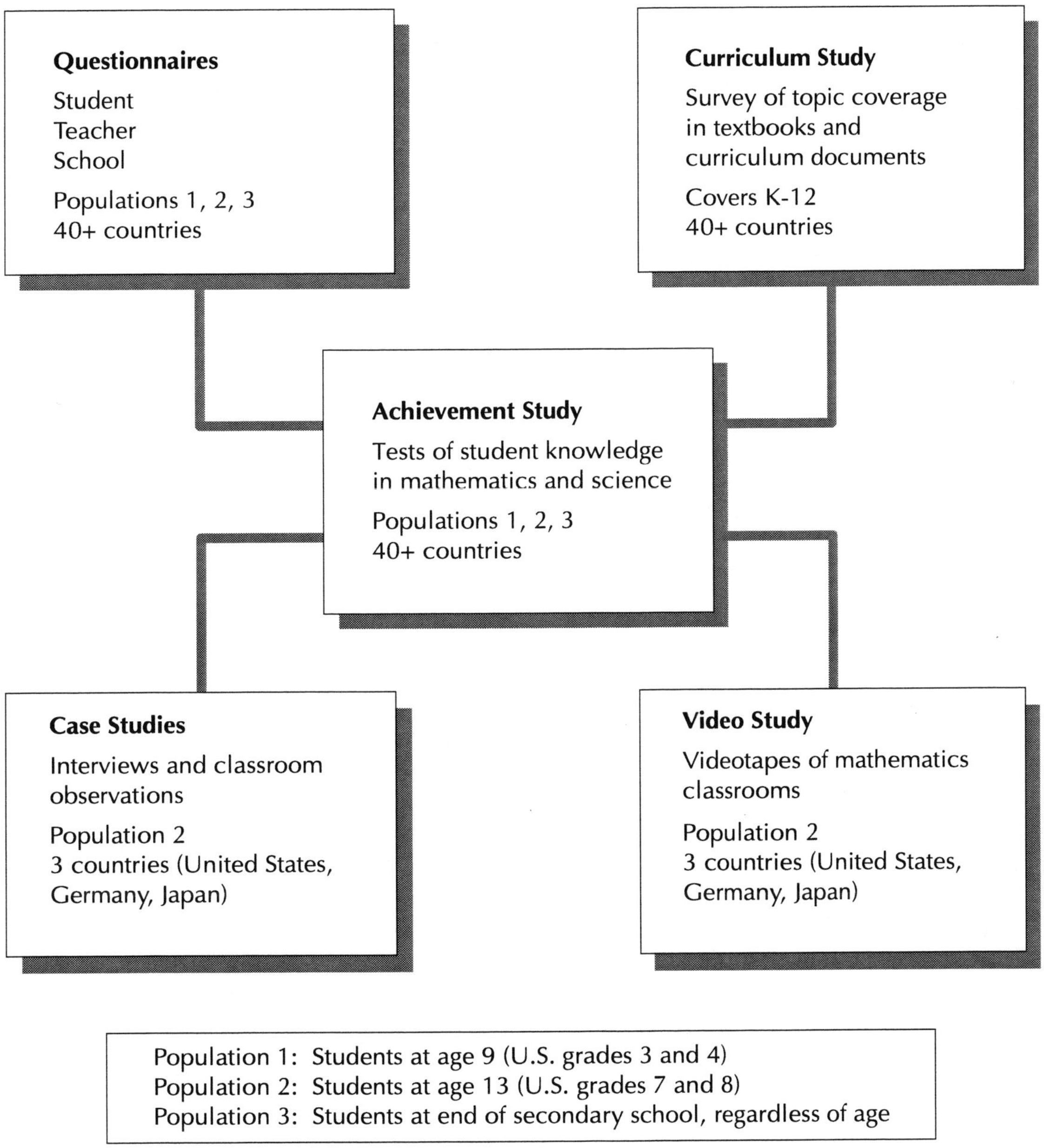

FIGURE 1 The basic structure of TIMSS.

been done, and second, that policy makers, practitioners, and others are drawing conclusions from analyses that may not be supported by further exploration of the data. Such conclusions, the board fears, are influencing both practical and policy decisions being made every day.

To stimulate discussion about the questions and issues arising from TIMSS that merit further exploration in the dataset, the board convened a workshop. Thirteen scholars, representing a range of interests, expertise, and seniority, as well as degrees of direct experience with the TIMSS data ranging from none to extensive, were asked to respond in written memoranda to a set of specific questions. These questions, which are discussed in detail below, were designed to assist the board in addressing specific concerns it had identified about establishing priorities for future research, identifying the kinds of knowledge claims best supported by TIMSS data, standards for the kinds of support such claims should have, and ways of combining different types of data. A number of other interested observers participated as well; all were given an opportunity in advance to review TIMSS documents as well as the memoranda from the scholars asked to respond to the questions BICSE had formulated. The workshop was then structured around a series of focused discussions, some with subsets of the group and some that included the entire group. These discussions were specifically designed to promote interdisciplinary thinking because of the unique breadth of TIMSS's design.

The board then deliberated on the results of the workshop discussions, tying them to the extensive discussions of TIMSS that the board itself has had during the past decade, and agreed on a set of recommendations that it intends as a useful guide and a spur to action for those eager to mine the dataset. These recommendations are the primary focus of this document, but they are not intended only to provide guidance to potential researchers and funders. Many groups—politicians, parents, math and science educators, and others—have paid attention to TIMSS and hope to use it to help students learn. BICSE is convinced that for TIMSS data to benefit students in this way, those who are using it need to understand both its complexity and its limitations.

ORIGINS: THE CONCEPTUAL BASIS FOR TIMSS AND THE BOARD'S ROLE

BICSE's particular interest in TIMSS grew out of an overall concern with the contribution that international comparative studies can make both to the intellectual study of education and to the discussions of policy choices that occur at all levels of the U.S. education system. The board framed its views of the value of international comparative studies, and of ways in which they could be improved, in a report published in 1993, *A Collaborative Agenda for Improving International Comparative Studies in Education* (National Research Council,

1993). The board returned to some of the issues identified in that report as it began to reflect on the contribution TIMSS has made and can make to education research and policy. One issue highlighted in the 1993 report has played a particularly significant role in TIMSS: the importance of considering the contexts in which learning takes place whenever comparisons are made. The board wrote:

> . . . comparisons of achievement levels are not meaningful unless one can, first, identify the educational inputs and processes that contribute to observed outcome differences between countries; second, make some estimate of the contribution of each educational input to realized outcome levels, and, third, consider how these effects vary by context. (National Research Council, 1993:12)

The board also outlined reasons why both qualitative studies and large-scale surveys are needed to provide complete and useful comparisons among countries. Qualitativc studics such as ethnographies, case studies, and others, the board noted, "allow us to understand what it means to be educated in diverse settings around the world." Such studies can provide the deep understanding of societies that makes it possible to interpret and explain the results obtained from large-scale studies. Large-scale achievement studies, in turn, provide "the only way to obtain a simple numerical comparison of a large number of countries on a common set of measures." In the case of TIMSS, the size of the study also provides the possibility of a variety of comparisons on a scale far beyond what an independent researcher could hope to address (National Research Council, 1993:22-23).

TIMSS was, of course, designed to produce, on a previously untried scale, the rich comparisons that a combination of qualitative and quantitative studies could generate. The study design was based on a conceptual model that was also used in the Second International Mathematics Study (SIMS), the first IEA study to explicitly address the contexts in which learning takes place. That model was described this way:

> The study was . . . conceptualized as an examination of mathematics curricula at three levels: the *intended curriculum* as transmitted by national or system level authorities; the *implemented curriculum* as interpreted and translated by teachers according to their experience and beliefs for particular classes; and the *attained curriculum*, that part of the intended curriculum learned by students which is manifested in their achievement and attitudes . . . . The curriculum at each of these levels is influenced by the context in which it occurs and the contexts themselves are determined by a number of antecedent conditions and factors.

Four general research questions, grounded in this model, were developed for TIMSS, and each component of the study was devised in response to one or more of these questions (see Robitaille and Garden, 1996:37-43):

- How do countries vary in the intended learning goals for mathematics and science; and what characteristics of educational systems, schools, and students influence the development of these goals?
- What opportunities are provided for students to learn mathematics and science; how do instructional practices in mathematics and science vary among nations; and what factors influence these variations?
- What mathematics and science concepts, processes, and attitudes have students learned; and what factors are linked to students' opportunity to learn?
- How are the intended, the implemented, and the attained curriculum related with respect to the contexts of education, the arrangements for teaching and learning, and the outcomes of the educational process? (Robitaille and Garden, 1996:37-43)

The study addressed three separate populations, students at age 9 (Population 1); students at age 13 (Population 2); and students, regardless of age, in their last year of secondary school (Population 3). TIMSS was designed to gather data about what mathematics and science the students had learned and augment it with information about their schools, teachers, textbooks, and curricula. It included a test of student achievement containing both multiple-choice and open-ended items, as well as performance items for some students; a set of questionnaires directed at students, teachers, and school administrators; a curriculum study that provides a comparative picture of standards and curricula in participating countries; a set of case studies of schooling in Germany, Japan, and the United States; and a videotape study of mathematics instruction for middle-school students in the same three countries.[4] It is important to note that different countries participated in different subsets of the study components—a point that is easily lost in discussions of the relationships among the components. Even this brief description of the study illustrates the complexity that resulted from the attempt to produce the context-embedded comparisons called for by the framework.

It was not the study designers' intention that these components of TIMSS stand alone, but rather that their results be used, and combined, to produce deep understanding of the factors that most influence learning. Although the intention that the contextual data should be used in this way is clear in the conceptual framework, the extent to which precise empirical links among particular components of the study were expected, and the means by which they could be established, were not articulated. The process by which questions about the relationships among the components can be pursued has not been completed—and may never be completed without conscious attention and support from both the research and funding communities.

At this phase of the study's progress, the board wanted to focus its

[4]See Appendix A for a more detailed description of the study components.

workshop on both learning from what has been accomplished and understanding what is involved in completing the comparisons the study makes possible. With these goals in mind, board members, led by Lynn Paine and Francisco Ramirez, planned a workshop to explore ways in which further analyses could best build on the work that has already been done. Their overarching goal was to find ways to identify research directions that would meet the dual criteria of being both approachable through the TIMSS data and of real importance to the education policy and practice communities.

To stimulate the desired conversations, it was necessary to include a wide variety of perspectives not only on TIMSS itself, but also on the standards by which evidence ought to be judged. Three sets of questions—see boxes on pages 8, 9, and 10—were used to focus a diverse group of scholars' attention on three fundamental elements of the context for learning addressed by the TIMSS conceptual framework: the effects of curriculum on achievement; the links among professional development, teaching, and achievement; and the factors that influence individual achievement. Brief memoranda prepared in advance about these sets of questions served as the starting point for the workshop. The goal for both the memoranda and the workshop itself was to stimulate ideas rather than to present carefully crafted arguments. The discussion ranged widely—from a number of focused discussions of possible future analyses to broad consideration of goals for TIMSS. Questions arose about the infrastructure that will be required to meet those goals, about some of the study's contributions and disappointments, and about the perspectives different disciplines bring to bear on the study as a whole and on its components.

What Does TIMSS Tell Us About Curriculum and Its Effects on Achievement?

Conclusions drawn from the curriculum component of TIMSS have perhaps received more sustained public attention than any other aspect of the study. The particular claim that has received the most attention—that cross-national variation in academic achievement is influenced by and accounted for by cross-national variation in the content of science and mathematics curricular frameworks—is an important one. TIMSS researchers have highlighted several different dimensions along which they found cross-national variation:

- the extent to which curriculum emphasizes critical thinking skills,
- the range of subject matter covered in the textbooks and teacher reports,
- the extent to which subject matter was well integrated within and across grade levels, and
- the existence of national standards and national curricular frameworks that contain content and performance expectations.

How might the claim that these factors account for variation in performance be assessed using the full range of data TIMSS has made available?

What Does TIMSS Tell Us About the Links Among Professional Development, Teaching, and Student Achievement?

No single substudy answers this question directly, yet virtually every component of TIMSS provides data that can help us explore this line of inquiry. In the teacher questionnaires, case studies, and videotapes of teaching, we get glimpses of teachers' preservice preparation, their professional development opportunities, instructional choices they make, and features of their classroom practice. How do we move across these various types of data? Which aspects of these questions is TIMSS well equipped to answer, and which are basically beyond the reach of TIMSS? More specifically,

- What difference does mentoring and a supported period of induction make to teachers? Both *Pursuing Excellence* and the TIMSS Resource Kit video note that these things do make a difference; how can we interpret the significance of this finding? How can other TIMSS data help?
- The statement has been made that teachers in Japan have more opportunities to discuss practice than do their colleagues in the Unites States. Is there sufficient data across the datasets to support this statement? If so, how ought we to interpret it?
- TIMSS researchers claim that Japanese teachers "widely practice what the U.S. mathematics reform recommends, while U.S. teachers do so less frequently" (*Pursuing Excellence*, eighth grade). What sort of variation is there in practice within country, across grades, across subjects? How might these variations influence our descriptions of practice?
- What is the link between professional training and support and teachers' willingness to teach subjects in particular ways?
- How can we use information from the teacher questionnaires about their teaching particular content in particular ways?

What Does TIMSS Tell Us About the Factors That Relate to or Influence Individual Achievement?

Obviously TIMSS was designed with the explicit goal of identifying factors that relate to achievement, and virtually every component of TIMSS provides data that can help us explore this line of inquiry. What is less clear is how to move across the various types of data. Decisions about which research questions to pursue should be based on a thoughtful balance between the importance of the possible answers and the relative power of the TIMSS data to shed light on the questions. Which aspects of these questions is TIMSS well equipped to answer, and which are basically beyond the reach of TIMSS? More specifically, how can the study help with these questions:

- Is the amount of race/class achievement variation in the U.S. larger than that in other countries, and, if so, what are the implications?
- What can be learned from comparing within-country variability across different levels of aggregation, such as class or school?
- When attempting to understand U.S. achievement relative to other countries, which of the achievement subscales are most useful and why?
- What clarifications for standard setting in the context of the National Assessment of Educational Progress (NAEP) can be obtained from the TIMSS results?
- Is U.S. achievement relative to that of other countries improving, staying the same, or slipping over time?

Recommendations for Making the Most of TIMSS

The major conclusion that BICSE has drawn from its observation of the progress of TIMSS thus far is that the study will be incomplete unless a variety of follow-up analyses of the data are undertaken.

Despite the undisputed value of the reports that have already been released, the total cost of TIMSS will be hard to justify if no more comes out of the study than those and others already scheduled for release. Moreover, while the reports that are part of the primary work of TIMSS were all reviewed, they are nevertheless official reports that have not received the kind of open peer review to which independent scholarship is generally subject. The board urges that each of the scholarly communities with an interest in TIMSS explore the hypotheses suggested by TIMSS, the data that have been collected, and the methodological issues the study has raised. This work will be important not only to scholars, but also to the teachers, administrators, and policy makers who need to draw inferences from TIMSS. Although much, but not all, of the existing data have been made available through a variety of channels, the board believes that the needed scholarship should be encouraged and facilitated in several ways.

This report identifies and describes the research approaches that BICSE has identified as the most promising ways of following up on the data from TIMSS. The recommendations outlined in this report address both practical issues surrounding further analyses and intellectual goals for following up on the work that has already been done. The workshop yielded a variety of specific suggestions for secondary analyses, as well as a predictably wide divergence of opinions about the relative merit of particular ones. Although the board did not identify individual research topics on which it would place top priority, it synthesized this discussion, weighed it together with information and reflections related to TIMSS that it has accumulated over a number of years, and developed specific criteria for identifying the approaches that it believes hold the most promise. The board believes—and this position was strongly endorsed at the workshop—that independent scholars should be encouraged to develop their own ideas and not be confined by rigid parameters of acceptable approaches to the data. The purpose of the board's recommendations is not to impede the creative process by which scholars explore possibilities and subject their ideas to the scrutiny of their peers. Rather, the board intends both to speak out about the importance of secondary analysis—additional research designed to follow up on the study's initial findings—and to offer some considered judgments about the possibilities the data hold. The recommendations are followed by some explanation and description of a few specific examples of approaches that illustrate how they might be applied; the board has not taken a position on the priority that should be assigned to any one of these research approaches.

RECOMMENDATION: FOCUS ON ANALYSES THAT TAKE ADVANTAGE OF WHAT IS UNIQUE ABOUT TIMSS

TIMSS is different from other studies in several specific ways, and for well thought-out reasons. Most obviously, TIMSS is big. It assessed the knowledge of students in nearly fifty countries and made significant compromises in the development and administration schedule, selection of the content to be studied, and other issues to do so. It addressed both mathematics and science learning, which entailed compromises in sampling and other difficulties.[5] It assessed intact classes of students. TIMSS was also, of course, complicated by the inclusion of the curriculum, videotape, and case studies. Each of these decisions multiplied the difficulty of interpreting results and linking findings, but they also made possible a number of approaches to the data. Thus far, little of the work that has been done exploits the complex design of the study, but there are many ways to do so. The board strongly urges that such work be undertaken.

Participants in the workshop spent a good deal of time discussing specific approaches to the data that would exploit the multiple-country, multiple-age, two-subject study design. The general approach that received the most attention is that of probing much more deeply into the cross-national comparisons, though a variety of ways of doing this are possible. As David Baker noted, it is the cross-national design that will "enable TIMSS to have its biggest impact on policy formation in the United States," (Baker, 1998:5), because this is what makes it possible to identify system-level factors that vary across countries—such as school governance, standards enforcement, textbook dominance, and many other features—and look at how they relate to differences in national performance.

One approach to this line of inquiry is the comparison of patterns of within-country variance. For example, Douglas Grouws noted that

[5]For example, for Population 2, researchers identified intact mathematics classes and administered assessments in both subjects to those students. Application of this method was complicated by the fact that students in a mathematics class do not always stay together for their science instruction, so science teachers were linked by other means. This method did make possible the linking of student, school, and system characteristics, and of mathematics and science results (Robitaille and Garden, 1996. 47-48).

TIMSS has shown that U.S. 8th graders spend more hours in mathematics classes than do students in Germany and Japan (Grouws, 1998:1-2). He pointed out that this information, though presumably accurate, is not sufficient reason to conclude that the number of hours spent in class did not contribute to differences in achievement (Japanese Population 2 students performed better than both U.S. and German Population 2 students). It may be, he explained, that in one or more of the countries there is extreme variance from school to school in the hours spent in class; such a finding could open up a new avenue for investigation and would also, of course, mitigate against policy recommendations based on the mean alone.

Richard Murnane made a similar point when he stressed the importance of comparing across nations the degree of variation in student achievement within nations, and tying this variation to measures of socioeconomic inequality (Murnane, 1998). Similarly, Elizabeth King advocated such comparisons of within-country variation because of what they can reveal about "the importance each country accords equality across groups or its success in achieving this equality" (King, 1998:3).[6] By going deeper than the achievement rankings in this way, they and others argued, researchers can make much more useful connections between TIMSS and the kinds of policy questions that are of most interest in the United States

Mary Metz and numerous others remarked on the value of cutting the data in different ways than has been done and comparing, within countries, subgroups of schools, classes, and students; subsets of the domain; performance on particular item types; etc. (Metz, 1998:7). Other areas of interest would include variations in opportunities to learn science at the three grade levels, particular characteristics (such as structure, format, content) of the mathematics and science items that were scaled together, and the relationship of the items to school curricula as perceived by teachers. Fleshing out the details of the performance of students within a particular country can help to answer a variety of questions about its system and about what its students are taught.

Other ways of pushing beyond the achievement rankings were discussed. Gender differences, for example, were observed across nations in the performance of Population 2 students but not in that of Population 1 students. This finding could make an important contribution to the debate about the underlying causes of gender differences in science and mathematics achievement but needs to be examined further through exploration of the specific topics that do and do not yield differences; changes in curriculum, teaching strategies, and other issues during the years from the elementary to middle school; and patterns of gender difference across countries.

---

[6]Elizabeth King prepared a written summary of her ideas in advance but was unable to attend the workshop.

Many participants also stressed the importance of comparing performance across mathematics and science. As Aaron Pallas noted,

> Since within a country the same group of students are being tested in math and science, different profiles of math and science achievement cannot be accounted for by differences in students' social backgrounds and educational resources, . . . there may be some merit to looking carefully at what is going on in countries that appear to be doing appreciably better in one subject area than the other (Pallas, 1998:4-5).

TIMSS was designed to provide opportunities for many different comparisons—using many different criteria but the same set of students—in the hope that the contextual factors with the greatest effects on student learning could be identified. More of the possible comparisons need to be made in order for TIMSS to be more than a collection of separate studies.

RECOMMENDATION: REVISIT THE CONCEPTUAL FRAMEWORK ON WHICH TIMSS WAS BASED

As has been discussed, the design of TIMSS grew out of a conceptual framework that embodies some propositions about the ways students learn. The framework lays out four basic research questions (cited earlier) to be addressed—about the role of the intended and implemented curricula in influencing learning, and about the attained curriculum. This framework reflects specific ideas about which factors affect learning and how they interact with one another (Schmidt, Jorde et al., 1996). Now that the data have been collected, it is important that researchers take time to reassess that framework and the assumptions to which it led in the design and development of the TIMSS study.

For example, the selection of questionnaire topics reflects a variety of assumptions about the role of school structure and many other "inputs" in student learning; the design of the curriculum study reflects some assumptions about the relationship between teaching and curricular exposure; decisions made in the course of video and case studies reflect implicit and explicit ideas about factors that are most likely to influence learning and how those might be explored. It is important now to look back and assess the extent to which the results fit the study model. A second step in such an evaluation would be to consider alternative framework models to determine whether others might actually fit the data better.

RECOMMENDATION: DRAW LESSONS FROM THE CONDUCT OF TIMSS

Because TIMSS was, in a sense, experimental in both its scale and its goal of integrating essentially disparate kinds of data, the board believes it is extremely important that the research and policy communities look closely for lessons TIMSS can teach about practical, methodological, and conceptual issues. The board identified three particular areas that offer lessons for the development and design of future studies.

Recommendation: Explore Methodological Issues

There is clearly a need for research that can yield not only challenges to or support for particular ideas about teaching and learning, but also useful insights about how such ideas can be tested and explored. Because TIMSS was methodologically experimental, it is important that the scholarly community learn all that it can from the experience. Numerous questions suggest themselves, including:

- How well did the cognitive items in the various subdomains work?
- What has been learned about the design and conduct of a classroom video study?
- How might one go about explicitly linking very different kinds of data?
- What are useful, creative alternatives when explicit links are not possible?

Ways of bridging the gaps between different components of the study—another obvious way of exploiting its unique design—generated considerable discussion. One important message that emerged was that moving across the datasets will be tricky. The workshop participants clearly held a wide range of opinions about the value of doing so and, indeed, about the relative usefulness of the different components of the study. Part of what these discussions revealed was how little opportunity researchers in different fields have to interact with one another in substantive ways and to engage with the issues pertinent in other fields. For example, participants with qualitative backgrounds were disinclined to think that the questionnaire data regarding professional development for teachers was of much use, while more quantitatively inclined participants were equally cautious about how the videotapes and the case studies could best be used. The conversations that pushed participants with different views to consider new perspectives made plain that simple resolutions for many of the issues are not likely.

There was, as Lynn Paine put it, "a great deal of discomfort" about the feasibility of many of the possible links among the substudies. Several participants for whom TIMSS was new remarked that the conceptual underpinnings of the study were not easily apparent in either the discussions they had witnessed or the results about which they had heard and read. In attempting to describe the essential problem, the group noted that the components of TIMSS seem to stand alone and that the means of establishing desired links among them are not easy to see. Moreover, it was clear that the hoped-for links may not all be possible.

The board's conclusion is that such links could be established only through careful further research conducted by interdisciplinary teams. Such work might seem to some to be a purely intellectual exercise. The board believes it will be integral not only to making the most of TIMSS, but also to moving comparative education research forward and enabling it to provide information that can be of real value for decision makers.

Recommendation: Use the Problems and Successes of TIMSS to Inform the Planning of Future Studies

Testing and exploration of the conceptual framework of TIMSS is an important step in the long-term intellectual inquiry of which TIMSS is a part. In more immediate terms, however, it is also very important that practical lessons with implications for future studies be distilled from TIMSS now, both so that problems that arose with TIMSS will not be replicated and so that its successes can be. One problem that was mentioned more than once at the workshop related to the sampling issues that entailed important trade-offs (discussed below). Follow-up analyses that explore the costs and benefits of these choices would be of great use to planners of future studies. Of the numerous other aspects of TIMSS that were significant challenges, and that could yield practical lessons, a few examples are: the instrument was translated into more than 30 languages; open-ended responses from students in more then 40 countries were scored; questionnaires were designed to obtain useful information from students, teachers, and administrators in vastly different contexts. The technical manual for TIMSS (Martin and Mullis, 1996) provides a lot of detail about many of the steps that were taken, but further objective exploration is needed. There may well be an increased demand for large-scale studies such as TIMSS in the future, but practical lessons learned from the present study can benefit more modest comparative studies as well.

Recommendation: Lay the Groundwork for Making the Most of Future Results

The value of TIMSS will clearly be enhanced if future data collection efforts build on the foundation TIMSS has laid. For example, TIMSS results indicate that, at least in the United States, there may be a decline in achievement relative to other countries between the 4th and 8th grades. This inference, however, is a tentative one since TIMSS compares two different populations of students (4th and 8th graders in 1995), not the same population at two different points in time. If researchers focus now on questions that could be followed up as cohorts mature or questions that could be pursued in more depth in future studies, those studies will be better able to take advantage of such opportunities. In addition, by obtaining a more detailed picture of the curriculum over those years and of the case study findings about those years, and following up on other relevant clues in the data, researchers can prepare for other important analyses. Future studies may also include countries that did not participate in TIMSS, and this is another area in which advance planning could be beneficial.

Plans are already in place for a replication of TIMSS for Population 2, known as TIMSS-R. TIMSS-R will be administered in 1999, when the students who were in Population 1 for the original TIMSS will have moved into Population 2. It will include administration of a refurbished set of TIMSS cognitive items,[7] slightly modified questionnaires, and a video study of mathematics and science classrooms. Approximately 40 countries will participate. The National Center for Education Statistics (NCES) also has plans to conduct an additional follow-up study in 2003, to collect data about the original Population 1 students as they prepare to leave secondary school. Work should be done now to prepare to make the most of these efforts, and others as well.

RECOMMENDATION: EVALUATE CLAIMS THAT HAVE ALREADY BEEN MADE BASED ON THE DATA

Role of Curriculum

One particular set of claims based on TIMSS data—that cross-national variation in academic achievement is influenced by and accounted for by cross-national variation in the content of curricular frameworks for science and mathematics, and that the U.S. students' weak performance can be attributed to a curriculum that is "a mile wide and an inch deep"—was the subject of special attention at the workshop.[8] BICSE members chose to address this claim in part because it is the most specific causal claim that has been put forth in TIMSS reports and in part because this claim has received a significant amount of public attention. Noting that the publications describing and supporting this claim have not been reviewed by their authors' academic peers (because of their status as official project reports), the board believes strongly that follow-up studies that will explore, test, and confirm or contradict the claim are needed. This belief was

[7]Some of the original TIMSS items have been publicly released; these have been replaced for TIMSS-R with new items designed to be similar to the old ones.

[8]This claim is laid out in *A Splintered Vision* (Schmidt, McKnight, and Raizen, 1996).

reinforced for the board by the marked differences of opinion in evidence at the workshop about not only the validity of the claims, but also the standards by which they ought to be evaluated. Similar issues would be likely to arise with any TIMSS-based claims, and BICSE's view is that other serious claims ought to be explored and tested as they emerge.

A number of participants agreed that the claim that there is a connection between what is taught and what is learned is reasonable, so far as it goes. The finding that intended curricula in the United States generally include a greater number of topics, more repetition and review, and less intellectual focus than do those in other countries was intriguing but not entirely persuasive to the group—not all were convinced that this description of the U.S. curriculum tells the whole story. The further claim that it is the lack of coherence in U.S. curricula—and within the network of assessments, standards, professional development for teachers, and other factors that affect learning—that sets the United States apart from other countries and actually accounts for the weaknesses U.S. students demonstrated on TIMSS received even greater challenges. This claim was viewed by many in the group as a hypothesis not adequately supported by the evidence that has been put forward thus far. Participants suggested a number of counter hypotheses that deserve to be examined.

Among these were that rapid changes in curriculum in some countries may have, in effect, distorted the comparative picture that seemed evident to the authors of *A Splintered Vision*. TIMSS, by necessity, provides a snapshot of conditions at a particular moment; conclusions about differences in complex educational structures may require a longer view. Others suggested that important characteristics of the various countries and education systems studied that did not show up in the curriculum study may play a significant role in achievement differences. Heinrich Mintrop pointed out, for example, that East European countries are currently in the process of rendering their systems far less coherent than they had been before the fall of the Berlin wall because of their perception that a too-rigid coherence had been detrimental to learning (Mintrop, 1998). This point led to the suggestion that a curriculum that is focused and coherent is not necessarily also of high quality. Aaron Pallas pointed out, for example, that the study does not address the question of resources, noting that ambitious standards and effective alignment among texts, tests, and other elements might exist in a given system without the resources necessary to meet the standards. Pallas also pointed out that "even in a country with a centralized education system, no student experiences the *national* implemented curriculum; rather, students are exposed to a *particular* implemented curriculum" (Pallas, 1998:3).

The causal link between the configuration of curriculum and student learning may not have been forged in the extant reports, but many saw the data from the curriculum study as the source of intriguing directions for future study. Most participants agreed that although the curriculum study has provided a valuable macro-level view, the data need to be probed at a much more detailed level. James Shymansky noted that counting the number of topics covered per country, as was done in the curriculum study on which the claim about coherence was based, did little to illuminate the quality of teaching and learning (Shymansky, 1998:2). Jeremy Kilpatrick explained that for the curriculum study to address so many countries and grade levels, it was necessary for the notion of topic coverage to be conceived in a fairly crude way. He argued strongly for further research that would look much more closely at the U.S. curriculum, specifically to see what it "might be attempting that is not captured by the TIMSS framework" (Kilpatrick, 1998:1). It would also be useful, he explained, to develop more descriptive pictures of the treatment of particular topics in the curricula studied and to look more closely at the relationship between specific coverage of a topic and details about student performance on it that could emerge from item analysis. More fine-grained analyses such as these would not only be valuable in themselves, but could also yield further evidence with which to evaluate the claim that the U.S. curriculum's shallowness has caused the relatively poor performance of U.S. students.

More detailed pictures of the education systems in a number of countries would also be of use in this context because, as a number of participants pointed out, the degree to which a coherent curriculum is in place in a given nation is only a small part of the story of how learning occurs there. Investigation of specific factors that affect curricula, such as standards, teacher qualifications, and general population characteristics, could significantly enrich understanding of both intended and implemented curricula. More detailed explorations of how curricula in a few nations are constructed, and of the roles of different actors in their respective systems, could be of significant value. Augmenting the existing data with information of this kind could help to support the claims that have been made, or offer new ways of looking at the curriculum data.

A number of participants noted that a variety of cultural factors must surely play a role in shaping student learning and that a better understanding of those factors will illuminate the apparent curricular differences that have been identified. Moreover, as Gary Natriello suggested, analyses that leave citizenship aside and group the sampled population according to characteristics of their educational experience might be very useful in exploring the ways the intended and implemented curricula interact and affect learning (Natriello, 1998:5). A number of participants shared the view that the reports of the curriculum study may not have distinguished adequately between exposure to curriculum and the instructional modes through which the curriculum was delivered. As Aaron Pallas noted, "It may well be that the effect of exposure to a particular topic on achievement is contingent on the nature of that exposure—that is, the context in which the exposure occurs, the specific classroom (and extramural) activities that are intended to enable students to learn the topic, even perhaps the sequencing of topics that both precede and follow a given topic" (Pallas, 1998:5).

The workshop discussion only reinforced the board's concern that it would be easy for policy makers and the public to deduce from the public discussion of the curriculum study that it has proved a set of assertions about the effects of curriculum that have not actually been proved. Follow-up research is needed to pursue the most intriguing implications of and questions about this data and to provide policy makers and practitioners with a more nuanced picture of the role of curriculum in student learning. Other claims that have been, or may be, made based on the TIMSS data require follow-up as well. For example, a 1998 U.S. Department of Education report contains the claim that "the TIMSS results indicate a pervasive and intolerable mediocrity in mathematics teaching and learning in the middle grades and beyond" (Silver, 1998:1). As Gerald LeTendre pointed out, claims such as these "are then countered by rhetorical invectives that move the dialogue . . . away from sustained academic debate" (LeTendre, 1998:1). In the board's view, academic debate about what TIMSS shows is precisely what is needed.

A Word About Some of the Criticisms of TIMSS

In the context of the need to evaluate claims about TIMSS, it is important to address some criticisms of the TIMSS data that have received public attention. These criticisms concern relatively complex technical issues, and the details of the critical debate need to be explicitly aired. Indeed, in the board's judgment, the criticisms embody hypotheses about data-based comparisons that ought to be evaluated along with other claims.

The primary criticism is that the Population 3 (students completing secondary schooling) results are misleading because the ages and numbers of years of schooling of the students in the sample vary so widely around the world. Certainly the Population 3 data are complicated. The sampling design for Populations 1 and 2 were relatively straightforward by comparison; the goal was to discover how much students had learned by two ages, so in each country, the two grades serving the largest numbers of 9- and 13-year-olds, respectively, were sampled. For Population 3, however, the issues were more complex for a number of reasons. First, different systems structure their secondary schooling in different ways and conclude it at different ages. Second, students generally have far more choice about their studies in the late secondary years than they do in the elementary or middle-school years, so defining the content to be tested for this population was more difficult than for the others. Moreover, students in many countries, including the United States, are placed in academic tracks that dictate the curriculum to which they will be exposed.

These and a host of other factors had to be considered when the designers of TIMSS determined which accessible data about secondary students would be of most value. The designers of TIMSS determined, collaboratively, that for Population 3, the most useful results would be for students "who are at the point of leaving school and entering the workforce or postsecondary education" (Mullis et al., 1998:13). Further decisions were made about the means of identifying subsets of this population who had done advanced work in physics and calculus. Critics have claimed that it is unfair to compare the performance of 21-year-old students, the age of students sampled in Iceland, with that of 17- or 18-year-olds, the age of students sampled in the United States and that the results are misleading (Bracey, 1998; Rotberg, 1998).

The board has three comments about this discussion. First, it is clear that the complexity of the issues surrounding the Population 3 data was well understood by the TIMSS researchers, as they are thoroughly documented in the report of the results (Mullis et al., 1998:11-28). That report discusses the problems the researchers faced and the decisions they made. Tables in the report present details about the cohort sampled, the domain covered, characteristics of each participating country, and many other issues. An appendix to the report supplies further details about the upper secondary system in each country, particularly details about the academic tracks in each. Although some critics of TIMSS have seemed to charge that the TIMSS researchers ignored or overlooked the issues with the Population 3 data, this is clearly not the case.[9]

Second, the criticisms of TIMSS may have (or already have had) the effect of undermining TIMSS as a whole, yet they relate almost exclusively to the Population 3 sampling. It is important to note that the sampling for Populations 1 and 2 was, as has been mentioned, more straightforward and has not been criticized on the same grounds.

Finally, the criticisms themselves need to be evaluated carefully. As David Baker pointed out (1998:7), "Most of the arguments that the study's critics make about the 'fairness' of the simple cross-national comparisons of achievement are really cross-national hypotheses about which national factors shape the process by which

[9]This issue is further complicated by criticisms of the study's predecessors, particularly the Second International Mathematics Study (SIMS), with regard to sampling. Critics of SIMS charged that, in the earlier study, some countries sampled only their ablest students while others assessed a broadly representative sample. The result, these critics argued, was rankings that were both unfair and inaccurate. Cognizant of these criticisms, the TIMSS designers addressed the issue early in the process. Having learned from SIMS that it would not be possible to ensure complete compliance with all sampling procedures in all countries, they made the existence of sampling compromises in particular countries very clear in the published reports and in the released data. The question of how rigorously the sampling rules were followed and documented is, of course, separate from the question of determining which secondary students should be sampled.

learning takes place in school." In other words, the decision to test students just completing whatever secondary schooling is offered in their country, regardless of their age, was made deliberately; therefore, what is needed is a debate to address the pros and cons of that decision rather than a dismissal of all that resulted from it. The challenge is actually not to the accuracy of the numbers, but rather to the premises underlying their collection and the inferences made on the basis of their analysis. These criticisms are really assertions that a different set of data ought to have been collected. The basis for these assertions is, presumably, a counter hypothesis about which factors affecting learning are most important. Such a hypothesis needs to be evaluated by the scholarly community just as other TIMSS-based claims do.

RECOMMENDATION: EXPLORE WAYS TO RELATE THE TIMSS DATA TO OTHER SOURCES OF DATA

Each country that participated in TIMSS has now received its own national data, and all have access to the international data. Anecdotal reports indicate that secondary research is already well under way in a number of countries, and it is likely that both direct collaboration with non-U.S. scholars as well as research that builds on existing work could considerably enhance the value to be gained from TIMSS. Data from other sources within the United States, such as the National Educational Longitudinal Study and the National Assessment of Educational Progress (NAEP), as well as other international data might also provide possibilities for interesting follow-up analyses. As is discussed elsewhere in this report, more detailed portraits of the education systems and contexts for learning in other countries could be of great value as researchers attempt to explore puzzles in the data and pursue specific comparisons between individual countries. Germany and Japan are obvious targets for future research because of the three-country studies that were part of TIMSS, but high-achieving countries such as Singapore, and others that stand out for various reasons, may be producing research that follows up on their own national data.

As secondary analysis moves forward, it will also be important for the TIMSS research community to find ways to take note of findings that confirm, complement, or contradict one another. Such findings can obviously prevent unnecessary duplication of effort; they can also strengthen policy conclusions and generate promising leads. Finally, it will be important for researchers to determine the extent to which the TIMSS data confirm or contradict existing sources of data that bear on mathematics and science learning at the three target age levels. Work has been done to establish statistical links between TIMSS and NAEP (U.S. Department of Education, National Center for Education Statistics, 1998). Because there are many differences between the two assessment systems, only limited formal links have been established, but the possibilities inherent in such links, and in looser connections with other sources of data, need to be explored. On a broader scale, the TIMSS results should be evaluated in light of findings from the considerable existing literature on mathematics and science learning.

RECOMMENDATION: CONSIDER OTHER APPROACHES TO THE ANALYSIS OF THE ACHIEVEMENT DATA

Researchers who did not participate in the creation of the achievement scores have suggested that alternative models might yield different results, and possibly more profound findings that bear on important questions about cognition and curriculum effects. Current total (and subtest) scores capitalized on what is common among all test items and ignore important parts of the test design that included three dimensions—(1) content areas, such as mechanics and magnetism; (2) performance expectations, such as recall, problem solving, and practical investigations; and (3) affective measures, such as attitudes and teacher and student beliefs—and an over-sampling of items in curricular areas commonly found across countries. Moreover, Kupermintz and Snow (1997) and Hamilton et al. (1997) have shown that mathematics and science achievement tests similar to the TIMSS tests are more complex than admitted by the current TIMSS scores[10] and that once the complexity is taken into account, these scores are sensitive to curriculum effects. Without in any way disparaging the results already produced by the TIMSS Study Center, the board strongly recommends that alternative models for creating achievement scores be explored by independent scholars. Given that the achievement results, and the rankings in particular, have been accorded a significant measure of political importance within the United States, it is vital that any additional insights these data might yield be mined and made public.

RECOMMENDATION: RECOGNIZE THAT TIMSS DATA CANNOT ANSWER EVERY IMPORTANT QUESTION

Although TIMSS is obviously a very rich dataset, it is, of course, not comprehensive. It is important that both the research and policy

[10]For example, the science achievement test included on the NELS:88 longitudinal survey can be empirically decomposed into components that measure (1) basic knowledge and reasoning, (2) quantitative science, and (3) spatial and mechanical ability. These components can be predicted, differentially, by student course-taking patterns.

communities recognize that TIMSS cannot provide an answer to every question; this is important to note not only because limited resources for research should not be wasted on lines of inquiry that are not promising, but also because recognition of the gaps in the TIMSS findings can spur valuable research targeted at questions about which information is needed.

An important example of this point is the relationship between the achievement scores and particular modes of teaching. Observers of TIMSS have noted that relatively little of the resulting discussion has seemed to focus on specific implications of the results for teaching; some discussion at the workshop specifically probed this point. Each component of the study attempted in its way to elicit information about the ways different aspects of teaching affect student learning. The questionnaires included a number of questions about teaching strategies, pedagogical beliefs, preparation and support for teachers, and other issues. The case studies explicitly probed aspects of the professional development, conditions of employment, and other factors in teachers' lives in three countries. The video study provides not only footage of lessons, but also coded analysis of different teacher actions and a questionnaire about the filmed teachers' views of the taped lessons. The curriculum study provides information about the official documents and textbooks that teachers use. Yet many who are professionally interested in teaching have concluded that although many of these results are of great interest, and provide a wealth of examples and ideas, they have not coalesced into clear, supportable conclusions about the effects different modes of teaching have on learning.

The questionnaires, for example, struck many at the workshop as providing only fairly limited information about teachers' preparation and practice. Some noted that because the questionnaire data relied on self-reports, their value was limited. Others noted that they provide very little detail about important questions; for example, a question about the highest level of education teachers have received does not address either the nature of the degree or the number of teachers' degrees that are in fields other than those in which they are teaching. Another question, about teachers' sources of information when creating lessons, provides a sense of how different kinds of resources are used, but, as Gerald LeTendre pointed out, one needs to understand the nature and role of these resources in particular settings to make much use of this data.

A number of specific questions one might hope to answer through TIMSS were posed for discussion at the workshop, and many of them seemed to many in the group difficult to address through the existing data. For example, Senta Raizen noted that "there simply isn't enough information available in any of the TIMSS datasets to differentiate sufficiently among countries either as to training and support received by teachers in each country or regarding teachers' willingness to teach in a particular way, let alone link these variables" (Raizen, 1998:7).

Other questions are more approachable through the data, many in the group noted. Many seemed to adopt Mavis Sanders's formulation that with regard to links among professional development, teaching, and student achievement, TIMSS data serves to "highlight topics . . . that require further exploration" and to "caution [us] against divorcing professional development, teaching, and student achievement from cultural context" (Sanders, 1998:1).

The videotapes provide compelling images of teachers at work in their classrooms for which a variety of uses have already been found. According to some, they even provide persuasive portraits of distinctly "Japanese," "German," and "American" teaching styles. (Further analysis is, of course, needed to ascertain whether the degree of variation in achievement between countries is actually higher than that within countries.) The tapes in particular, but presumably other TIMSS material as well, have provided practitioners and others with extremely valuable views of possibilities they may not have considered and new ways of addressing common challenges. Such information is an important benefit of TIMSS, but does not provide answers to broad questions about the relationship between teaching practice and student achievement. Similar limitations in what the case study data reveals about teacher practice were also discussed, but the point of the discussion was decidedly not to criticize work that has been done. Rather, the discussion, and the board's concern, was focused on the risk in attempting to derive from TIMSS insights it cannot empirically support. Questions that are of particular concern to many—about successful practice, the impact of the National Council of Teachers of Mathematics standards and other reforms, the efficacy of different models of teacher induction and support—may need to await further research.

RECOMMENDATION: A CLEARINGHOUSE TO PROVIDE RESOURCES AND SUPPORT FOR SCHOLARS IS NEEDED

The board has been following the release of the data thus far very closely. Many of the TIMSS researchers have assisted the board in its efforts to keep track of the data that have been collected and to understand the structures of the various datasets. The board greatly appreciates this assistance and has taken note of the efforts that have been made by the TIMSS Study Center staff, the National Center for Education Statistics, and others to release the data as quickly as possible and to make datatapes, users' guides, and the like available on the web and elsewhere; to hold training sessions; and to field questions. The efforts that have been made by all involved to disseminate the data quickly are also appreciated. Nevertheless, because TIMSS is so large and complex, the board believes there is a need for an infrastructure that will facilitate access to the data. Although there are a variety of steps that could be of help to scholars, BICSE believes that the funding of a clearinghouse for TIMSS information would be the simplest and most effective one.

A number of workshop participants described specific reasons why it may be difficult for independent scholars to use the TIMSS data. First, it is clear that even for those who have paid close attention to TIMSS, it has been difficult to follow the relationships among the various portions of data that have been collected and made available; currently there is no one source of comprehensive, detailed information about this. The curriculum study, for example, includes data for varying numbers of countries, some that cover kindergarten through the end of secondary school, some that are specific to the age populations targeted by TIMSS, and so forth. Workshop discussions about possibilities for linking one part of the study to another made clear that it is not efficient for independent scholars to, for example, spend time tracking down correspondence among different sets of data; this is work that, once done, should be made easily available to anyone considering possibilities for secondary analysis.

It is also clear that both relatively straightforward issues of translation, as well as more potentially significant questions about cultural context—the precise meaning of the words in different languages for "homework," for example—could be a hindrance to many scholars. A clearinghouse for information accumulated about such issues as interpretation of particular vocabulary used in the translation of a questionnaire item or detailed information about educational structures in TIMSS countries would be of great value. The opportunities for networking and taking advantage of what is already known would be not only a convenience for many researchers, but also a means of amplifying the value of each contributing piece of research.

A further obstacle for many researchers is the extremely complex design, particularly of the achievement study. Jane Hannaway, a workshop participant, voiced a concern echoed by many at the workshop when she said that "the set-up costs of getting up and running with TIMSS are exceedingly high," and recommended that ways be found to "lower the entry barriers." Many capable researchers would need guidance in traversing a dataset of this kind—a source for answers to specific questions about the BIB (balanced incomplete block)-spiraled design, conditioned and unconditioned variables, and the like.

A variety of other questions that were aired further illustrate the need for a centralized source of information. Each nation, for example, had the option of excluding particular items from its questionnaires or including additional ones for its own purposes. Moreover, as has been noted, secondary research is already under way in a number of TIMSS countries. Clearly independent scholars would benefit a great deal from access to up-to-date information on TIMSS-related data as it becomes available; possibilities for collaboration would increase, and many other kinds of information could be shared.

The provision of technical support for individual scholars, established procedures for dealing with particular complexities in the data, and facilitated coordination and communication among individual scholars and scholarly communities about methodological as well as conceptual issues (by means of a website for example) would be of significant value to the research community.

RECOMMENDATION: INDEPENDENT SCHOLARS OR TEAMS OF SCHOLARS SHOULD RECEIVE FUNDING TO CONDUCT SECONDARY ANALYSIS

As this report has attempted to make clear, the board would like to see a broad array of scholars from a variety of disciplines use the TIMSS data in a variety of ways. Much of the funding for such work will need to come from the institutions that have already funded much of TIMSS and from others with significant resources for education research. Such funding might go to teams of investigators, individuals, or both, but it is important to note that the establishment of a few dedicated teams of investigators would by no means preclude the conduct of a variety of field-initiated studies or the ordinary process of peer review. In recommending that funders consider their responsibility to ensure that the work of TIMSS is completed, the board does not mean to suggest that other work, funded through other means, is any less important. The board's primary concern is that without funding, much of the secondary analysis that is a crucial part of a large-scale study such as TIMSS will not be done.

Summary

BICSE has identified three distinct reasons why the data from TIMSS are valuable. First, teachers, administrators, and others directly involved in schooling can benefit not only from any definitive empirical findings about teaching methods, teacher development, curriculum structure, and many other issues, but also from simply seeing examples of the many alternative approaches to teaching, curriculum design, and the like, that their counterparts around the world have devised. Second, on a somewhat broader scale, policy makers are clearly in search of generalizable empirical findings that can assist them in making difficult decisions and addressing problems. Finally, the pursuit of knowledge about the structure, theory, and practice of education will have benefits in both the short and long term, not only by satisfying the curiosity of scholars—and generating new questions for them to ponder—but by providing the basis for the empirical findings and innovations that fuel progress in education reform.

But, as this report was designed to demonstrate, the work of TIMSS is not complete. The possible benefits of TIMSS will not accrue automatically. TIMSS is complex for important reasons, and consequently easy to misunderstand. Because of its multidisciplinary design, it offers the possibility of comparative insights deeply rooted in the differing contexts that affect student learning. If the tantalizing links among its components can be realized, it offers the possibility of moving comparative research methodology forward in significant ways. These goals can be reached only through further analyses of the data, and such research costs money. The board is surprised that funds have not already been earmarked for secondary research; the examples of many other countries in which secondary research is already well under way only reinforce the point that the United States lags behind. The research is needed not only to obtain further benefits from TIMSS, but also to keep it from being used in ways that are misleading. TIMSS was a vast, collaborative effort; further collaboration is needed to reap its full benefits.

This report has described the approaches to follow-up research based on TIMSS that BICSE believes have the most potential value, and has provided examples of possible specific directions that such research might take. The premises underlying the specific recommendations can be summarized in three simple points:

- Further research is needed in order both to rigorously evaluate existing knowledge claims and to generate and evaluate other hypotheses.
- Further research is needed to explore in detail the possibilities for links among the different studies that make up TIMSS.

- Further funding is needed to make the realization of the goals described in the first two recommendations possible. This funding should be directed toward two targets. First, it should facilitate the release of TIMSS datasets in easily accessible formats to encourage secondary data analysis. Second, it should support centers or teams of investigators committed to undertaking research that follows up on the data from TIMSS.

The board is eager to see the TIMSS data continue to benefit both the research and policy communities and hopes, through these recommendations, to contribute to this process.

References

Baker, David P.

1998 Thoughts on Analysis Possibilities for TIMSS. Memorandum commissioned by the Board on International Comparative Studies in Education. Available from the Department of Education Policy Studies, Pennsylvania State University.

Bracey, Gerald W.

1996 International comparisons and the condition of American education. *Educational Researcher* 25(1):5-11.

1998 TIMSS, rhymes with 'dims,' as in 'witted.' *Phi Delta Kappan* May:686-687.

Grouws, Douglas A.

1998 Observations About TIMSS. Memorandum commissioned by the Board on International Comparative Studies in Education. Available from the Department of Curriculum and Instruction, University of Iowa.

Hamilton, Laura S., E. Michael Nussbaum, and Richard E. Snow

1997 Enhancing the validity and usefulness of large-scale educational assessments: IV. NELS:88 science achievement to 12th grade. *American Educational Research Journal* 34(1)(Spring):151-173.

Kilpatrick, Jeremy

1998 Next Steps for TIMSS Regarding the U.S. School Mathematics Curriculum and Its Effects on Achievement. Memorandum commissioned by the Board on International Comparative Studies in Education. Available from the Department of Mathematics Education, University of Georgia.

King, Elizabeth M.

1998 Statement for Topic 3 Group. Memorandum commissioned by the Board on International Comparative Studies in Education. Available from The World Bank, Washington, D.C.

Kuppermintz, Haggai, and Richard E. Snow

1997 Enhancing the validity and usefulness of large-scale educational assessments: II. NELS:88 mathematics achievement to 12th grade. *American Educational Research Journal* 34(1)(Spring):124-150.

LeTendre, Gerald K.

1998 Overall Evaluation of the TIMSS Database. Memorandum commissioned by the Board on International Comparative Studies in Education. Available from the Department of Education Policy Studies, Pennsylvania State University, University Park.

Martin, Michael O., and Ina V.S. Mullis, eds.

1996 *Third International Mathematics and Science Study: Quality Assurance in Data Collection.* Center for the Study of Testing, Evaluation, and Educational Policy. Chestnut Hill, Mass. Boston College.

Metz, Mary Haywood

1998 Memo on TIMSS: Question #2—Teachers. Memorandum commissioned by the Board on International Comparative Studies in Education. Available from the Department of Educational Policy Studies, University of Wisconsin.

Mintrop, Heinrich

1998 Discussion on TIMSS Group 1: Curriculum. Memorandum commissioned by the Board on International Comparative Studies in Education. Available from the Education Policy, Administration, and Planning Department, University of Maryland.

Mullis, Ina V.S., Michael O. Martin, Albert E. Beaton, Eugenio J. Gonzalez, Dana L. Kelly, and Teresa A. Smith

1998 *Mathematics and Science Achievement in the Final Year of Secondary School: IEA's Third International Mathematics and Science Study (TIMSS).* Center for the Study of Testing, Evaluation, and Educational Policy. Chestnut Hill, Mass.: Boston College.

Murnane, Richard J.

1998 What Does TIMSS Tell Us About the Factors That Relate to or Influence Individual Achievement? Memorandum commissioned by the Board on International Comparative Studies in Education. Available from the Graduate School of Education, Harvard University.

National Research Council

1993 *A Collaborative Agenda for Improving International Comparative Studies in Education.* Board on International Comparative Studies in Education, Dorothy M. Gilford, ed. Washington, D.C.: National Academy Press. Available electronically: http://www.nap.edu/.

1997 *Learning From TIMSS: Results of the Third International Mathematics and Science Study. Summary of a Symposium.* Board on International Comparative Studies in Education, Alexandra Beatty, ed. Washington, D.C.: National Academy Press. Available electronically: http://www.nap.edu/.

Natriello, Gary

1998 Using TIMSS for Examining Curriculum and Its Effects. Memorandum commissioned by the Board on International Comparative Studies in Education. Available from the Department of Philosophy and the Social Sciences, Columbia University Teachers College.

Pallas, Aaron M.

1998 Memo on Topic 1. Memorandum commissioned by the Board on International Comparative Studies in Education. Available from the Department of Counseling, Educational Psychology and Special Education, College of Education, Michigan State University.

Raizen, Senta A.

1998 Linking Teachers and Teaching to Student Achievement. Memorandum commissioned by the Board on International Comparative Studies in Education. Available from National Center for Improving Science Education, Washington, D.C.

Robitaille, David F., and Robert A. Garden, eds.

1996 *Research Questions & Study Design.* TIMSS Monograph No. 2. David F. Robitaille, gen. ed. Vancouver, Canada: Pacific Educational Press.

Rotberg, Iris C.

1990 I never promised you first place. *Phi Delta Kappan* 72(4):296-303.

1998 Interpretation of international test score comparisons. *Science* 280(May):1030-1031.

Sanders, Mavis G.

1998 What Does TIMSS Tell Us About the Links Among Professional Development, Teaching and Student Achievement? Memorandum commissioned by the Board on International Comparative Studies in Education. Available from the Center for Research on the Education of Students Placed at Risk, Johns Hopkins University, Baltimore, Md.

Schmidt, William H., Doris Jorde, Leland S. Cogan, Emile Barrier, Ignacio Gonzalo, Urs Moser, Katsuhiko Shimizu, Toshio Sawada, Gilbert A. Valverde, Curtis McKnight, Richard S. Prawat, David E. Wiley, Senta A. Raizen, Edward D. Britton, and Richard G. Wolfe

1996 *Characterizing Pedagogical Flow: An Investigation of Mathematics and Science Teaching in Six Countries.* Boston: Kluwer Academic Publishers.

Schmidt, William H., Curtis C. McKnight, and Senta A. Raizen

1996 *A Splintered Vision: An Investigation of U.S. Science and Mathematics Education,* Boston: Kluwer Academic Publishers.

Shymansky, James A.
1998 Question 3: What Does TIMSS Tell Us About Factors That Relate to or Influence Individual Achievement? Memorandum commissioned by the Board on International Comparative Studies in Education. Available from the Regional Institute for Science Education, University of Missouri, St. Louis.

Silver, Edward A.
1998 *Improving Mathematics in Middle School: Lessons from TIMSS and Related Research.* Report #4331HA7 10013. Washington, D.C.: U.S. Department of Education. Available electronically: http://www.ed.gov/inits/Math/silver.htm/

U.S. Department of Education, National Center for Education Statistics
1998 Linking the National Assessment of Educational Progress (NAEP) and the Third International Mathematics and Science Study (TIMSS): Eighth-Grade Results. May. NCES 98-500. U.S. Government Printing Office, Washington, D.C.. Available electronically: http://nces.ed.gov/pubs98/linking98/.

Appendix A

Road Map: Descriptions of TIMSS Databases

This summary of current information about the data and reports from the Third International Mathematics and Science Study (TIMSS), an international comparative study conducted by the International Association for the Evaluation of Educational Achievement (IEA), was prepared by National Research Council (NRC) staff for use by participants in a workshop sponsored by the Board on International Comparative Studies in Education. It is based in part on presentations made to the board by the TIMSS researchers and in part on publicly available information.

OVERVIEW OF THE STUDY

The Third International Mathematics and Science Study is, as its name suggests, the third in a series of international assessments of the mathematics and science learning of students around the world that has been carried out by the IEA. This complex study, which was conducted in 1995 and 1996, tested an unprecedentedly large number of students and countries worldwide; it also included a set of substudies designed to provide a wide variety of contextual information that could support and illuminate the achievement results.

The data have now been collected, and the primary analyses have been completed. Most of the data is now or very soon will be available to researchers, but it exists in many forms and locations. This appendix provides descriptions of each set of data, and details about how researchers can gain access to them. Most of the information contained here is also available through the many websites and publications devoted to TIMSS that are already available; however, because of the large scope of TIMSS, there is no one source for information about each of its many facets. This appendix outlines the main purpose and nature of each of the study components; more in-depth understanding can be obtained from the publications and sources listed.

THE ACHIEVEMENT RESULTS

International Data

Overview

The international achievement data is the core of TIMSS. Students in three populations were tested. Population 1 included students enrolled in the two adjacent grades that contained the largest proportion of 9-year-old students at the time of testing, grades 3 and 4 in most countries; Population 2 included students enrolled in the two adjacent grades that contained the largest proportion of 13-year-old students at the time of testing, grades 7 and 8 in most countries; and Population 3 included students enrolled in their final year of secondary education.

All countries that participated in any aspect of TIMSS were asked to take part in the mathematics and science assessment of their Population 2 students. Many chose to participate in the Population 1 and Population 3 assessments as well. The assessments contain a combination of multiple-choice and open-ended cognitive items covering the range of mathematics and science topics designated in the TIMSS framework. This framework was developed through an international consensus process and represents the content on which a majority of countries could agree, rather than coherent curricula for either subject.

The database includes separate proficiency scale scores for mathematics and science, students' responses to each cognitive item, students' responses to performance items, and coding reliability data for cognitive and performance items. The data disks provide sampling weights and information on how to link teachers to students. They also contain programs that allow the user to convert the raw data files (in ASCII format) into SPSS system files or SAS data sets, to estimate sampling variance using the jackknife repeated replication method, and to convert item response codes to score values. Codebooks documenting the structure of the data files and the source, format, descriptive labels, and response option codes for all variables are available. Finally, the disks contain data almanacs—text files that display unweighted summary statistics for each participating country for each variable in the background questionnaires.

What Do the Data Consist Of?

Results from Achievement Tests in Mathematics and Science Administered to Sampled Students in All Three Populations in All Participating Countries All countries that chose to participate in any part of TIMSS were required to test their Population 2 students; each had the option of testing students in Populations 1 and 3. Countries also had the option of administering additional questions of their own to their own students and could also identify subsets of the

TIMSS items that were most relevant to the material their own students have been taught.

Population 1 Randomly sampled intact classes of students took the test, with each individual completing only a subset of the questions prepared for this population. There were 102 math items and 97 science items. In some countries a subset of students selected for the general assessment were randomly selected to complete a set of performance items. These sets presented students with four or five of twelve different extended (30-minute) problems; students used kits of materials to conduct experiments or complete other tasks.

Population 2 Randomly sampled intact classes of students took the test, with each individual completing only a subset of the questions prepared for this population. There were 151 math items and 135 science items. In some countries a subset of students selected for the general assessment were randomly selected to complete a set of performance items, as above.

Population 3 Randomly sampled intact classes of students took the test, with each individual completing only a subset of the math and science questions prepared for this population. No hands-on performance tasks were administered to Population 3 students. Countries also had the option of identifying students in two "specialist" populations—students enrolled in either an advanced mathematics course or a physics course—who were tested for their knowledge in these two subjects. This option makes it possible to compare results for advanced students from all of the countries who chose this option.

Responses to Questionnaires Administered to Students, Teachers, and School Administrators Basic questionnaires were administered in all participating countries; each country had the option of including additional questions for their own purposes in their questionnaire booklets. Some questions were covered in more than one set of questionnaires so that responses of students, teachers, and administrators with regard to particular issues could be compared. Responses to each of the question types can be linked, so that, for example, students' perceptions of their classroom activities can be compared to those of their teachers. It should be noted that not all questionnaire items yielded useful results.

Student Questionnaires Questionnaires administered to all sampled students addressed topics such as classroom and out-of-school activities, home background, and attitudes toward mathematics and science. Some demographic data, such as the students' gender, home language, and parents' education level is also included. Additional questions about items found in the home and leisure activities fill out the picture of students' lives somewhat. The questionnaires for *Populations*

1 and 3 include specific questions about the instruction students have received in mathematics and science, such as how new topics are introduced and how frequently they use calculators, computers, and worksheets. For *Population 2*, there are separate versions of the questionnaire for mathematics and science students, which probe some of the details of their instruction in these areas.

Teacher Questionnaires The mathematics and science teachers of sampled students in *Populations 1 and 2* were asked questions about issues such as their background, education, and professional training; their views about mathematics and science; their responsibilities within the school; and the support they receive. A number of questions addressed teachers' coverage of specific topics in mathematics and science and their pedagogical approaches both in teaching certain material and in their general handling of students. Separate versions of the questionnaire were administered to mathematics and science teachers of *Population 2* students to probe details about their classroom strategies and topic coverage; one version was administered to all teachers of *Population 1* students, which contained some of these kinds of questions. No *Population 3* teacher questionnaires were administered because it was not possible to link students and teachers at this level.

School Questionnaires Administrators at the schools of sampled students were asked to respond to questions about school staffing and resources; responsibility for major activities such as determining course content, assigning teachers to classes, and establishing discipline policies; mathematics and science course offerings; and support for teachers. Other questions address the nature of the school's student population and problems the school confronts, as well as remedial and enrichment opportunities it offers. The questionnaires for *Populations 1 and 2* are very similar; the *Population 3* school questionnaire includes some additional questions about staffing for advanced courses, tracking, graduation requirements, and other issues.

Results of the Test Curriculum Matching Analysis This analysis was designed to assist users of the data in understanding how well the items in the mathematics and science tests correspond to material taught within the participating countries. Experts within each country reviewed copies of each item and judged the extent to which it was appropriate for that country's students. Each country's performance on just the subset of items identified as appropriate for its students was then computed. In addition, each other country's performance on that subset of items was computed, so that results can be compared.

Publicly Released Items Approximately three-fifths of the total items administered will be released to the public. Those not released will be available for future testing so that performance trends over time can be established. Item characteristic statistics are included for each released item.

How Can One Gain Access to the Data?

Data for all three populations have been released. The international achievement data was prepared by the staff at the TIMSS International Study Center at Boston College.[1]

The Database, User's Guide, and Released Items All can be ordered from Boston College, and are also available on the International Study Center's web page (http://wwwcsteep.bc.edu/TIMSS1.html).

Compact Disk: Data Tapes Data for each population is on a separate disk. Files are in ASCII format. CDs can be obtained from the IEA Secretariat.

What Has Been Published?

A user's guide to the data is available both in printed form and on the World Wide Web. The guide describes TIMSS, including the data collection instruments, sample design, and data collection procedures. It also documents the content and format of the data files and provides example analyses. The guide provides guidance as to appropriate uses of the different files and variables included in the data. Four supplements to the guide contain copies of the background questionnaires as well as documentation of the ways in which particular questions were adapted for use in particular countries. Documentation of derived variables included in the published international reports is also included. The user's guide can be ordered from IEA or downloaded from the Boston College web address above.

A number of reports based on the international database have already been released:

Mathematics and Science Achievement in the Final Year of Secondary School: IEA's Third International Mathematics and Science Study

Mathematics Achievement in the Middle School Years: IEA's Third International Mathematics and Science Study

Science Achievement in the Middle School Years: IEA's Third International Mathematics and Science Study

Mathematics Achievement in the Primary School Years: IEA's Third International Mathematics and Science Study

Science Achievement in the Primary School Years: IEA's Third International Mathematics and Science Study

[1] All addresses are listed at the end of this appendix.

Performance Assessment in IEA's Third International Mathematics and Science Study

Pursuing Excellence: A Study of U.S. Twelfth-Grade Mathematics and Science Achievement in International Context

Pursuing Excellence: A Study of U.S. Eighth-Grade Mathematics and Science Teaching, Learning, Curriculum, and Achievement in International Context

Pursuing Excellence: A Study of U.S. Fourth-Grade Mathematics and Science Achievement in International Context

Individual Country Data

Overview

The international database is organized by population and by country. Any data for each of the participating countries beyond what is included in the international dataset can be obtained through the national research coordinator for the individual country. Some participating countries embedded data collection instruments of their own in the TIMSS material; some have conducted their own analyses of the some of the data that has been released. Names and addresses of coordinators are listed in the Acknowledgments sections of the achievement reports. The coordinator for the United States is William Schmidt.

The U.S. data was collected by WESTAT, which has prepared U.S. datafiles to supplement the international files. The files can be used with the National Center for Education Statistics' (NCES) Electronic Codebook, which means that researchers who want to combine international questionnaire variables with U.S. variables will be able to do so relatively easily. Further information about U.S. TIMSS can be obtained from the U.S. TIMSS website at http://ustimss.msu.edu/.

What Do the Data Consist Of?

Results for the United States of Background Questions Administered to the International Population A few of the questions on the international questionnaire were not administered to the U.S. sample, such as those concerning tracking, for example.

Results of Background Questions Administered to Students, Teachers, and Administrators in the U.S. Sample Only ("National Variables") These results provide data on students' race and ethnicity, details about students' home background and study habits, teacher background, classroom practices, whether schools are public or private, and other variables. The datafiles will be set up in the same manner as the international files, and guidance for linking student and teacher files will be provided.

Special Sampling Weights, Appropriate for Analyzing Data on the Teacher Questionnaire in the United States These weights were developed to handle missing data; procedures limiting the response burden on teachers in the United States resulted in gaps in the teacher data for the United States.

How Can One Gain Access to the U.S. Questionnaire Data?

User's guide and data almanacs are available from NCES.

The Results of the Curriculum Study

Overview

The curriculum study was designed to explore differences in the structure and content of mathematics and science instruction in the TIMSS countries. It was an examination of both textbooks and curriculum documents in each subject and was supplemented by interviews with curriculum experts within the participating countries. The study was lead by William Schmidt.

What Do the Data Consist Of?

The data on which the curriculum study was based consist of a variety of curriculum materials and textbooks collected by the team. These materials were coded and analyzed in various ways. Below is a description of what is available.

Topic Trace Mapping Experts within each country used national and, in some cases, regional curriculum documents as the basis for coding their country's coverage of the mathematics and science topics included in the TIMSS framework.[2] The result is a visual representation of the extent to which each country covers each of the individual topics (44 in mathematics and 82 in science) at 12 different age/grade levels. The codes represent various categories of coverage that are possible for a particular country at a particular level, such as "topic already covered in previous grade, not covered this year," "topic introduced this year," and the like. The coding was done by panels of experts from within each country. This data covers 46 nations.

Document Analysis—Data about the Curriculum Guides and Textbooks This dataset is the result of the research team's analysis and

[2]The United States posed a particular difficulty for the researchers because of its policy of allowing states and districts to make choices about curriculum. The research team used an amalgam of six frameworks to come up with the U.S. codes.

coding of documents relevant to the three age levels tested in TIMSS. The documents in this analysis were selected to represent those used with at least half of the students in the targeted grades; the researchers worked with country experts to select the documents. The coding was done by native-language speakers within each country; coders were trained in all of the coding categories, and reliability checks were done. The dataset is in three parts that are coded in such a way that they can be linked to one another.

Document-Level File This file lists each of the documents that were analyzed by the research team and provides information such as its publication date, the grade levels for which it was intended, and other basic data. Approximately 1,600 documents, both texts and guides for both subjects, are included.

Unit-Level File The researchers identified sets of material within the textbooks that could be considered units, defined as representing 1 to 3 days of instruction. This file identifies the units and provides basic information about them, such as what they cover, how long they are, and how they are presented. (This analysis was done for texts only, not curriculum guides.)

Block-Level File This file presents the most detailed information about the content of the textbooks. Blocks of content—these are conceptual sections smaller than the units described above—were identified and coded. Information such as the primary and secondary content area covered in the block, number of pages devoted to the block, and primary and secondary levels of performance expectation are provided. Individual content blocks in all textbooks (not curriculum guides) were coded. Using this file, it would be possible for a researcher to identify particular pages of individual textbooks in the collection—pages that cover a particular topic, for example—and locate those pages in the actual book at the Michigan State Study Center.

Manuals These list each of the variables used and define the codes.

The Textbooks and Curriculum Guides All of the materials collected by the research team are housed at the TIMSS Study Center at Michigan State University, organized by country and by grade level. The researchers did their initial coding of these documents by marking directly in these books, so researchers can see what was done. Visitors are welcome to use these materials at the site, and staff are available to assist them.

Summarized Data for Individual Countries Several countries have requested their own curriculum data, and the research team has made it available not only in its raw form, but also in an intermediate, summarized form that is easier to use. This is available to all participating countries.

How Can One Gain Access to the Curriculum Study Data?

The curriculum study data tapes will be available from the Michigan State Study Center.

What Has Been Published?

A Splintered Vision: An Investigation of U.S. Science and Mathematics Education

Many Visions, Many Aims: A Cross-National Investigation of Curricular Intentions in School Mathematics

Many Visions, Many Aims: A Cross-National Investigation of Curricular Intentions in School Science

A Summary of Facing the Consequences: Using TIMSS for a Closer Look at United States Mathematics and Science Education

Facing the Consequences: Using TIMSS for a Closer Look at United States Mathematics and Science Education

The Results of the Video Study

Overview

The video study was designed to provide a window into middle-school mathematics classrooms in three countries, Germany, Japan, and the United States. The research team, led by James Stigler, identified a subset of the classrooms that were sampled in the Population 2 portion of the main TIMSS assessment and videotaped them in their entirety. The taped classrooms were chosen to be representative of 8th-grade classrooms in the three countries. Because the sample of classrooms was a subset of the testing sample, it may be possible to link the U.S. videos to the achievement results. This is probably not possible for the results from Japan because in some cases different classrooms were substituted for taping in that country; similar technical difficulties with the German sample would also make linking to achievement results difficult.[3] Researchers interested in linking the results may wish to consult WESTAT or the TIMSS Study Center at Boston College for assistance.

What Do the Data Consist Of?

Several kinds of data were produced by the videotape study.

---

[3]The source for this information about linking is Keith Rust of WESTAT.

The Videotapes One complete lesson in each of 231 different classrooms across the three participating countries was taped; there were 81 U.S. classrooms, 100 German, and 50 Japanese. The tapes have been converted to digital IMPEG1 files so they can be viewed and scanned on a computer.

Transcripts All of the lessons that were taped have been translated into English and transcribed (this was a single process so only English-language versions are available). The transcripts include running time codes that make it possible to refer to individual lines or blocks of dialogue, and an indication of who is speaking (teacher or student). They were originally prepared as a rough guide for navigating through the tapes; some notes and pictures have been added subsequently to make them more useful to someone who does not have access to the tapes, but this process has not been comprehensive. The transcripts are Microsoft Word files that can be searched for keywords.

Tables The research team has produced tables for the entire set of lessons, which provide descriptions of both the mathematical content of each lesson and the pedagogical activities that occurred during the lesson. These are linked to the time codes on the transcripts, and are intended to be used as tools for finding things in the transcripts and for understanding what is happening in each lesson. The descriptions of the content include representations of diagrams or other visual aids used by the teacher and other relevant notes. The content was described by mathematicians who reviewed the tapes.

Supplementary Images The research team has collected additional materials that are relevant to particular lessons, such as pages from textbooks and worksheets. These have also been digitally scanned.

Codes and Categories All of the videotaped lessons have been coded for various events, teaching strategies, and content elements. Some codes indicate the frequency of certain events; others indicate the duration of various activities. The code development team established a standard of 80% interrater reliability. These codes are available in an SPSS file.

Manuals A transcription manual and a videographers manual were prepared for the study.

Public-Use Videos Because of the need to maintain the privacy of those who were videotaped for the study, the tapes cannot be released to the public. However, many people were very eager to see the tapes, so the research team made a set of six tapes that replicated the conditions in which the study tapes were made. The teachers and students involved in these new tapes were volunteers who gave permission for the tapes to be made public. There are two tapes from each of the three countries. The taped classrooms were chosen to reflect what the researchers believed was typical of each country, not the range of competence within any country.

Results of the Questionnaire A questionnaire was administered to the teachers of each of the classrooms that were taped (this questionnaire is entirely separate from the teacher questionnaire that was part of the main TIMSS battery). Through a combination of open-ended and multiple-choice questions, it explored in detail the teachers' views of the lesson that was taped—its content, their goals for it, strategies used during it, and their sense of how typical it was. Additional questions cover their reaction to the taping itself and their exposure to current thinking about teaching and learning mathematics.

How Can One Gain Access to the Videotape Study Data?

The rules governing access are different for the different kinds of data described above.

The Videotapes Because it would be impossible to disguise the identity of the students and teachers who were videotaped, access to the tapes is restricted. The tapes are housed at the TIMSS Video Research Center at UCLA, which has a site license. Researchers may use them there if they sign a confidentiality agreement. The center provides the necessary hardware and software, as well as native Japanese and German speakers to assist visiting researchers. Researchers may also apply for their own site licenses and obtain copies of the data. It will be possible to purchase the entire dataset or a subset (90 lessons—15 algebra and 15 geometry, randomly selected for each country). The digitized tapes will all be on a single hard-disk server at the UCLA center so the entire database can be searched at once.

Everything Else All of the material described above that is not restricted will be available on a single CD ROM, which can be obtained from the U.S. Department of Education's National Center for Education Statistics (NCES). The videotapes prepared for public use are available from the Superintendent of Documents (GPO #065-000-01025-9).

What Has Been Published?

A formal report on the results of the videotape study has not yet been published, though several articles describing aspects of it have been.

The Results of the Case Studies

Overview

The case studies were designed to provide contextual information about the experiences of students and teachers in three countries, Germany, Japan, and the United States. Harold Stevenson, the principal investigator for the study, has described it as "a descriptive study . . . a description of what you would find if you were in these particular cultures." He also described it as a sort of hybrid between the methods of anthropological ethnography and those of psychology; the study was a combination of interviews with individual students, parents, and teachers, and observations of classroom lessons in grades 4, 8, and 12. The product is a combination of interview transcripts and summaries and field notes made during and after the observations. The research was collected by a team of 19 researchers who were both fluent in the necessary language and skilled in interviewing and observation techniques.

The case study research focused on four major research topics:

- Standards: What kinds of standards exist in this culture? Where and how are they developed? How are they implemented? How do they affect teaching and learning?
- Individual Differences: What do people see as the causes of ability differences? How are ability differences dealt with? What are the practices for educating extremely disabled and extremely gifted students?
- The Role of Secondary School in Adolescents' Lives: How do students spend their time both in and outside of school? What are students' attitudes toward school? What influences and pressures exist in their lives?
- The Working Lives of Teachers: How and why do people become teachers? How do they spend their time both in and outside of school? What opportunities exist for professional development? How do they feel about their work? What are their general working conditions?

What Do the Data Consist Of?

Case study observations and interviews were conducted at three sites in each country, one primary site (in each case, the largest of the selected cities) and two secondary sites. The sites were selected in consultation with country representatives; the selection was designed to yield sites comparable in "size, geographic distribution, and economic base." None of the sites was rural. Four researchers (one for each topic) went to the primary site in each country for 2 to 3 months; for each of the secondary sites, two researchers spent 2 weeks. Researchers visited elementary, middle, and secondary schools. All classroom observations took place in 4th-, 8th-, and 12th-grade classrooms. In each case they attempted to identify a mix of high- and low-achieving schools. Selection of schools, classrooms, and parents was done through consultation with local officials. The research was conducted in 1995.

The case study database consists of word processing files (Microsoft Word and WordPerfect) organized by country, by topic, and by researcher. It is not coded but can be searched by keywords (sets of keywords for each topic were identified in advance). The data all fit on a CD ROM. Text is labeled as being either interview or observation data.

- Transcriptions and summaries of interviews with students, teachers, administrators, and parents; a total of 600 interviews were conducted.
- Observations (field notes) of lessons and general observations of schools and neighborhoods; 200 separate school and classroom observations were conducted.

How Can One Gain Access to the Case Study Data?

Access to the case study database is restricted because it was impossible for the researchers to disguise the identities of those who were willing to be interviewed and observed sufficiently to protect their privacy. Consequently, researchers who would like to use the database must apply for site licenses and sign confidentiality agreements. Those who would like to gain access should call Cynthia Barton of the National Center for Education Statistics at 202/219-2199 for details.

What Has Been Published?

A five-volume report of the case study results is being published, to be available on the U.S. Department of Education's home page on the Internet (http://www.ed.gov), and in alternate formats upon request. The five volumes, listed below, consist of a separate report for each of the three countries, a synthesis report that integrates the four research topics for all three countries, and a summary of the research that was done in advance to prepare for the field work. Further information about these publications can be obtained at the website http://www.ed.gov or Internet address TIMSS@ed.gov/.

The Educational System in Germany: Case Study Findings
The Educational System in Japan: Case Study Findings
The Educational System in the United States: Case Study Findings
To Sum It Up: Case Studies of Education in Germany, Japan, and the United States
Contemporary Research in the United States, Germany, and Japan on Five Education Issues

OTHER PUBLICATIONS

The following publications are available from Pacific Educational Press:

Curriculum Frameworks for Mathematics and Science, 1993. TIMSS Monograph Series No. 1

Research Questions and Study Design, 1996. TIMSS Monograph Series No. 2

Mathematics Textbooks: A Comparative Study of Grade 8 Texts, 1995. TIMSS Monograph Series No. 3

National Contexts for Mathematics and Science Education: An Encyclopedia of the Education Systems Participating in TIMSS

Quality Assurance in Data Collection

Technical Report No. 1 on Design and Development

RESOURCES

International Association for the Evaluation of
Educational Achievement (IEA) Secretariat
Herengracht 487
1017 BT Amsterdam
The Netherlands
telephone: +31 20 625 3625
facsimile: +31 20 420 7136
e-mail: Department@IEA.nl
IEA Clearinghouse on the World Wide Web: http://uttou2.to.utwente.nl/

Kluwer Academic Publishers Group, Order Department
P.O. Box 358, Accord Station
Hingham, MA 02138-0358
telephone: 617/871-6600
facsimile: 617/871-6528
e-mail: kluwer@wkap.com
World Wide Web: gopher://gopher.wkap.nl

National Center for Education Statistics
TIMSS Project
555 New Jersey Ave., N.W., Suite #402A
Washington, DC 20208
telephone: 202/219-1333 (TIMSS Customer Service Line)
facsimile: 202/219-1736
e-mail: TIMSS@ed.gov
World Wide Web: http://www.ed.gov/NCES/timss

Pacific Educational Press
Faculty of Education, University of British Columbia

University of British Columbia
Vancouver, B.C., Canada V6T 1Z4
telephone: 604-822-5385
facsimile: 604-822-6603
e-mail: cedwards@interchange.ubc.ca
World Wide Web: http://www.curricstudies.educ.ubc.ca/projects/TIMSS/

TIMSS International Study Center
Center for the Study of Testing, Evaluation, and Educational Policy
Campion Hall 323
School of Education, Boston College
Chestnut Hill, MA 02167
telephone: 617/552-4521
facsimile: 617/552-8419
World Wide Web: wwwcsteep.bc.edu/timss

TIMSS U.S. National Research Center
College of Education
457 Erickson Hall
Michigan State University
East Lansing, MI 48824-1034
telephone: 517/353-7755
facsimile: 517/432-1727
e-mail: jbabcock@pilot.msu.edu
World Wide Web: http://ustimss.msu.edu

TIMSS Videotape Study Center
UCLA Department of Psychology
Graduate Mail Room
Los Angeles, CA 90095-1653
telephone: 310/794-9554
e-mail: jj@ucla.edu

U.S. Superintendent of Documents
P.O. Box 371954
Pittsburgh, PA 15250-7954
telephone: 202/512-1800
facsimile: 202/512-2250

Attaining Excellence: A TIMSS Resource Kit. Available from the U.S. Government Printing Office and on the World Wide Web at http://www.ed.gov/NCES/timss

Resources on the World Wide Web related to TIMSS:

- http://ra.terc.edu/alliance/TEMPLATE/regional_networks/cia/assessment/timss.cfm
- TIMSS Video Study: General Discussion http://forum.swarthmore.edu/epigone/vstudy

Appendix B

Workshop Agenda and Participants

NEXT STEPS FOR TIMSS: A BICSE WORKSHOP ON SECONDARY ANALYSIS

June 17-18, 1998
National Academy of Sciences/National Research Council
2001 Wisconsin Avenue, Washington, D.C.

AGENDA

Day 1: June 17

8:00 a.m.	*Continental breakfast in meeting room*
8:30-9:00	***Welcome and introductions*** Brief discussion of goals for the workshop and agenda Lynn Paine and Francisco Ramirez
9:00-12:00	***Detailed exploration of selected research topics*** Three topic teams, each led by a BICSE member, work in separate rooms to address the questions raised about each topic and to identify key areas for further discussion.
12:00	*Lunch in meeting room*
1:00-2:00 p.m.	***Whole group assessment of team discussions*** Rapporteurs from each group will report on the discussions and on the key areas for further discussion identified for their respective topics.
2:00-5:00	***Beginning a synthesis*** Three cross-disciplinary teams will be formed, each containing a few representatives from each of the three topic teams. These new groups will build on the morning's identification of key issues in each topic area. The purpose of this reshuffling of the group is to ensure that insights gained from consideration of one set of questions will be applied to others. These sessions will focus specifically on two issues:

• Do the claims warranted from one study find confirmation in others?
• Are new claims made possible when different kinds of data are brought together?

Overnight — Rapporteurs will be asked to prepare a summary of the discussions and report to the whole group the following morning.

Day 2: June 18

8:00 a.m. — *Continental breakfast in meeting room*

8:30-10:30 — ***Debriefing and moving forward***
Whole group will meet to discuss the results of the previous day's discussion. Rapporteurs will report on cross-disciplinary teams' conclusions, and the BICSE leaders will lead the whole group in:

- identification of consensus
- identification of divergent views
- discussion of implications for establishing research priorities

10:30-12:15 — ***Synthesizing the discussion***
Lynn Paine and Francisco Ramirez will lead the group in a discussion of lessons to be drawn from the workshop discussion. Key areas to be addressed will include:

- establishing priorities for future research
- identification of the kinds of knowledge claims best supported by TIMSS data
- standards for the kinds of support knowledge claims should have
- suggestions about ways of combining different kinds of data

12:15-12:30 — ***Closing remarks***
Lynn Paine and Francisco Ramirez

12:30 — ***Adjournment***

PARTICIPANTS

David P. Baker, Department of Education Policy Studies, Pennsylvania State University

Bennett I. Bertenthal, U.S. National Science Foundation, Arlington, Virginia

***Christopher T. Cross,** Council for Basic Education, Washington, D.C.

***John A. Dossey**, Department of Mathematics, Illinois State University

Pascale D. Forgione, Jr., National Center for Education Statistics, Washington, D.C.

Michael Garet, Pelavin Research Institute, Washington, D.C.

Eugenio Gonzalez, TIMSS International Study Center, Boston College

Douglas Grouws, Department of Curriculum and Instruction, University of Iowa

Jane Hannaway, The Urban Institute, Washington, D.C.

Stephen P. Heyneman, The World Bank, Washington, D.C.

Pamela Jakwerth, American Institutes for Research, Palo Alto, California

Harry George Judge, Brasnose College, University of Oxford

Takako Kawanaka, Department of Psychology, University of California, Los Angeles

A. Eamonn Kelly, U.S. National Science Foundation, Arlington, Virginia

Jeremy Kilpatrick, Department of Mathematics Education, University of Georgia

***Michael W. Kirst**, School of Education, Stanford University

Daniel Koretz, The RAND Corporation, Washington, D.C.

***Paul G. LeMahieu**, University of Delaware and Delaware Department of Education

Gerald LeTendre, Department of Education Policy Studies, Pennsylvania State University, University Park

Marlaine E. Lockheed, The World Bank, Washington, D.C.

Mary Haywood Metz, Department of Educational Policy Studies, University of Wisconsin

***Mary M. Lindquist**, Columbus State University, Columbus, Georgia

Heinrich Mintrop, Education Policy, Administration, and Planning Department, University of Maryland

Ina V.S. Mullis, TIMSS International Study Center, Boston College

Richard Murnane, Graduate School of Education, Harvard University

Gary Natriello, Department of Philosophy and the Social Sciences, Columbia University Teachers College

---

* Member, BICSE

David Nohara, Rensselaer, New York
Martin E. Orland, National Center for Education Statistics, Washington, D.C.
Eugene Owen, National Center for Education Statistics, Washington, D.C.
***Lynn W. Paine**, Department of Teacher Education, Michigan State University
Aaron Pallas, Department of Teacher Education, Michigan State University
***Andrew C. Porter**, School of Education, University of Wisconsin, Madison
Senta A. Raizen, National Center for Improving Science Education, Washington, D.C.
***Francisco O. Ramirez**, School of Education, Stanford University
Mavis G. Sanders, Center for Research on the Education of Students Placed at Risk, Johns Hopkins University
William H. Schmidt, U.S. TIMSS National Research Center, Michigan State University
Joel Sherman, Pelavin Research Institute, Washington, D.C.
James Shymansky, Regional Institute for Science Education, University of Missouri, St. Louis
Larry E. Suter, U.S. National Science Foundation, Arlington, Virginia
Ineko Tsuchida, Developmental Studies Center, Oakland, California
Trevor Williams, Westat, Inc., Rockville, Maryland
Kenneth I. Wolpin, Institute for Economic Research, University of Pennsylvania

Staff, Commission on Behavioral and Social Sciences and Education

Alexandra Beatty, Program Officer, Board on International Comparative Studies in Education
Michael J. Feuer, Director, Board on Testing and Assessment
Patricia L. Morison, Director, Board on International Comparative Studies in Education
Marie Suizzo, Program Officer, Division on Education, Labor, and Human Performance
Barbara Boyle Torrey, Executive Director